WILDERNESS ADVENTURES

Fishing

Seth Kingston

PowerKiDS press

NEW YORK

Published in 2022 by The Rosen Publishing Group, Inc.
29 East 21st Street, New York, NY 10010

First Edition

Portions of this work were originally authored by Suzanne Slade and published as *Let's Go Fishing*. All new material in this edition was authored by Seth Kingston.

Editor: Caitie McAneney
Book Design: Michael Flynn

Photo Credits: Cover (woman fishing) MoMo Productions/DigitalVision/Getty Images; cover, pp. 17, 19, 23, 29 (mountain logo) doyz86/Shutterstock.com; interior (mountain background) Lidiia/Shutterstock.com; p. 4 FedBul/Shutterstock.com; p. 5 John Fedele/Getty Images; p. 6 Min C. Chiu/Shutterstock.com; p. 7 Matthew Leete/DigitalVision/Getty Images; p. 9 SDI Productions/E+/Getty Images; p. 11 Nosyrevy/iStock/Getty Images; p. 12 Valery Ambartsumian/iStock/Getty Images; p. 13 IronHeart/Moment/Getty Images; p. 15 Jodie Griggs/The Image Bank/Getty Images; p. 17 Don Mason/Getty Images; p. 19 stock_colors/E+/Getty Images; p. 20 georgeclerk/E+/Getty Images; p. 21 JMichl/E+/Getty Images; p. 23 scottlitt/Shutterstock.com; p. 24 Piter1977/iStock/Getty Images; p. 25 Zocha_K/E+/Getty Images; p. 27 Michael Blann/DigitalVision/Getty Images; p. 29 Fertnig/E+/Getty Images.

Library of Congress Cataloging-in-Publication Data

Names: Kingston, Seth, author.
Title: Fishing / Seth Kingston.
Description: New York : PowerKids Press, [2022] | Series: Wilderness adventures | Includes index.
Identifiers: LCCN 2020039413 | ISBN 9781725329317 (library binding) | ISBN 9781725329294 (paperback) | ISBN 9781725329300 (6 pack)
Subjects: LCSH: Fishing--Juvenile literature.
Classification: LCC SH445 .K46 2022 | DDC 639.2--dc23
LC record available at https://lccn.loc.gov/2020039413

Manufactured in the United States of America

Some of the images in this book illustrate individuals who are models. The depictions do not imply actual situations or events.

Note to Parents and Teachers: Some of the activities in this book require adult supervision. Please talk with your child or student before allowing them to proceed with any wilderness activities.

CPSIA Compliance Information: Batch #CWPK22. For further information contact Rosen Publishing, New York, New York at 1-800-237-9932.

CONTENTS

Fishing—Then and Now

People have been fishing since **prehistoric** times, using their hands, nets, spears, and fishing rods. Images from ancient civilizations show us that people used fishing rods in ancient China, Egypt, and Greece. The first anglers, or fishermen, fished for food.

fish finder

Today, people still fish for food, but they also fish for fun. Many enjoy the challenges they must face to find, **attract**, and catch fish. Fishing is a great way to get outside and enjoy nature. People of any age, size, and skill can try fishing. You can buy lots of gear for your trip, from expensive rods to wireless fish finders that use **sonar**. Or, you can use a simple fishing rod and a worm from your backyard. Fishing is for everyone!

Fishing can be a fun outdoor family activity all year round.

Find a Fishing Spot

Before setting out into the wild, you should always make a plan. First, you need to choose a place to fish. Most creeks, rivers, lakes, and ponds make good fishing holes. Make sure to fish only in public areas that are **designated** for fishing. Look out for signs that say NO FISHING or PRIVATE PROPERTY.

Once you arrive at your fishing location, you need to find some fish. Watch for clues to help you find where fish might be hiding. Some fish swim near weed beds or reeds to hide from predators. Others can be found in areas of swirling water called eddies. Fish often stay in shaded water below a tree, pier, or docked boat. Spotting tiny bubbles can tell you there might be a fish underwater!

WILD FILE

When fishing in a river, look for fish near fast-moving water, such as below a dam or waterfall.

It's good to be still, quiet, and patient when you're looking for a fish.

What Should You Pack?

You don't need fancy gear to go on a beginner's fishing trip, but it's still smart to pack supplies. Think about how long you plan to be out. A fishing trip can last for hours, so take plenty of food and drinks. Wear a hat and sunscreen to keep from getting burned, and sunglasses to protect your eyes as you look for fish in the bright sun. Pack a life jacket that is comfortable and fits properly.

Your most important tool is your fishing rod, or pole. Connected to your pole is a reel. A reel is a device that holds a spool of fishing line. Tie a hook on the end of your line. Add a metal sinker above it to help your hook and bait sink in the water.

WILD FILE

If you're looking to catch fish closer to the water's surface, you might add a hollow ball called a float, or bobber, to your line. It'll keep the hook from sinking to the bottom.

Some people use a net while fishing. You can pack one along with your fishing pole!

Pack Your Tackle Box

Anglers usually keep their fishing supplies in a tackle box. A tackle box can hold your smaller fishing equipment and anything else you might need. Extra supplies like lines, hooks, and floats are important if your line breaks, a fish steals a hook, or your float drifts away.

You can bring different kinds of baits or lures, and different sizes of hooks for different fish. It's always smart to bring a first-aid kit in your tackle box and any tools you might need to cut a line or remove hooks from fish.

Bring a sturdy net to land your fish and something to hold your catch. Some anglers store their fish in baskets, buckets, or on a rope.

You can use a tackle box to organize your fishing gear into certain sections.

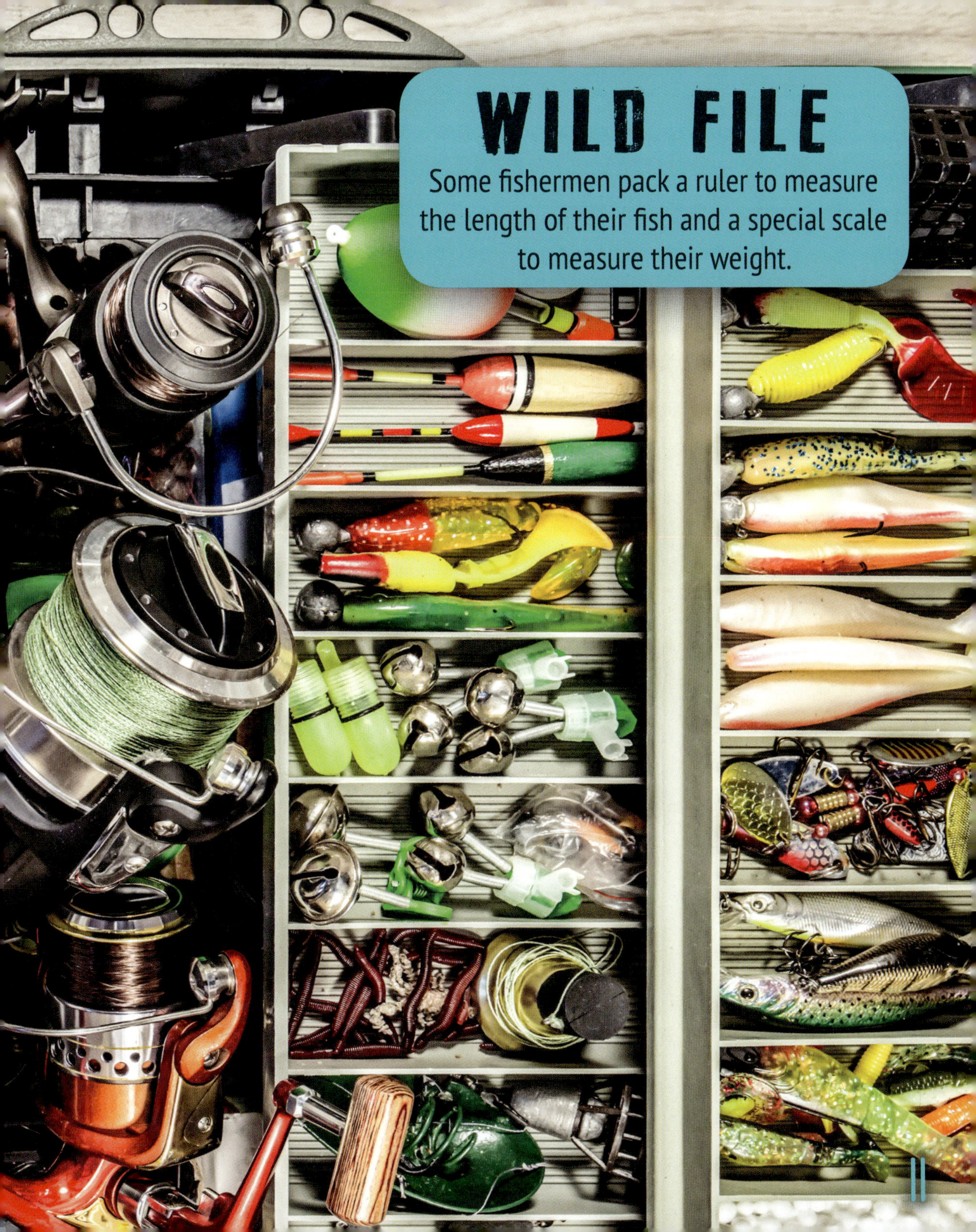

WILD FILE

Some fishermen pack a ruler to measure the length of their fish and a special scale to measure their weight.

The Right Bait

Bait is any material used to attract and catch fish. What kind of bait should you pack? It depends what kind of fish you'd like to catch. Big, wiggly worms are favorites when you're looking to catch bass or trout. Dough balls and bread are great bait for carp and whitefish.

You can use bait you find at home, such as corn, cooked pasta, cheese cubes, and bread balls. Earthworms from your yard also make great bait. Tackle shops sell many kinds of bait, such as red worms, night crawlers, wax worms, and maggots. If you are fishing for large fish, you can buy tiny fish called minnows.

Lures are non-food bait. Spinnerbait has spinning blades that **mimic** the motion of prey in water. Buzzbait creates **vibrations** to attract fish.

Some fishing lures look like small, colorful fish, which can attract bigger fish.

Cast Away!

You've got your fishing spot and your tackle box. You've put bait on your hook. Now it's time to catch some fish. Cast away!

Casting is the skill of throwing your hook and bait out into the water. There are several kinds of casts, such as the overhead cast and sidearm cast. Many anglers use an overhead cast. First, let out enough fishing line on your rod, bring it overhead and behind you, and cast it in front of you into the water. Before you start, check your surroundings for wires, trees, and people. Sidearm casting—or casting forward at a side angle—is useful when there are a lot of low-hanging trees or wires around. It helps to watch an experienced angler cast their line before you try it.

WILD FILE

Some people cast from a dock or boat. Others cast while standing in the water or on the shore. Choose your cast based on the conditions around you.

Surf fishing, or surf casting, is when you catch fish while standing on a beach or shoreline.

Fishing in Fresh Water

Fresh water is water that isn't salty. Lakes, rivers, ponds, streams, and creeks are all great fishing spots, full of freshwater creatures. You can fish from the shore, stand in shallow water, or fish off a boat. Some freshwater fish are small enough to hold in your hand, while the white sturgeon can grow to 20 feet (6.1 m) long.

Some freshwater lakes only allow catch-and-release fishing. That means you put your fish back into the water right after you catch it. This type of fishing protects lakes from running out of fish, while allowing anglers to enjoy the sport of fishing. When trying catch-and-release fishing, use a barbless hook, which doesn't have a sharp point at the end. Barbless hooks are less likely to hurt or kill the fish you catch.

HUGE FRESHWATER FISH

Many freshwater anglers in North America are interested in catching smaller fish like bass, trout, walleye, and salmon. But extreme anglers have caught some *huge* freshwater fish all over the world. The largest freshwater fish in the world may be the Mekong giant catfish, found in the Mekong River in Southeast Asia. It can weigh more than 600 pounds (272.2 kg)! Other freshwater giants include bull sharks, giant freshwater sting rays, arapaima, alligator gars, and paddlefish.

Many state and national parks offer freshwater fishing in lakes.

Out on the Ocean

Ocean fishing is a whole new challenge for anglers. Whether you go deep-sea fishing on a boat or fish from the beach, you can catch some amazing saltwater fish.

Some people fish from the shore. Fish often swim in deeper waters, so you'll have to cast your line far from the shore. You will know a fish is on the line when your float jerks below the water's surface. When a fish bites, pull your pole quickly to set the hook. Keep your fishing line tight as you reel in your catch. When the fish arrives at the edge of the water, wait for the next wave to wash it up on the beach. You'll find different saltwater fish in different regions. Look at a local fishing guide to identify your catch.

OCEAN PRIZE

Freshwater fish might grow big, but when you compare sizes, saltwater fish blow them out of the water! From a variety of shark species to huge swordfish and marlins, experienced anglers can try to catch some of the biggest fish in the sea. One of the largest fish you might catch in the ocean is a bluefin tuna. An adult bluefin can weigh more than 1,000 pounds (454 kg). Sometimes it takes anglers hours to reel in their prize!

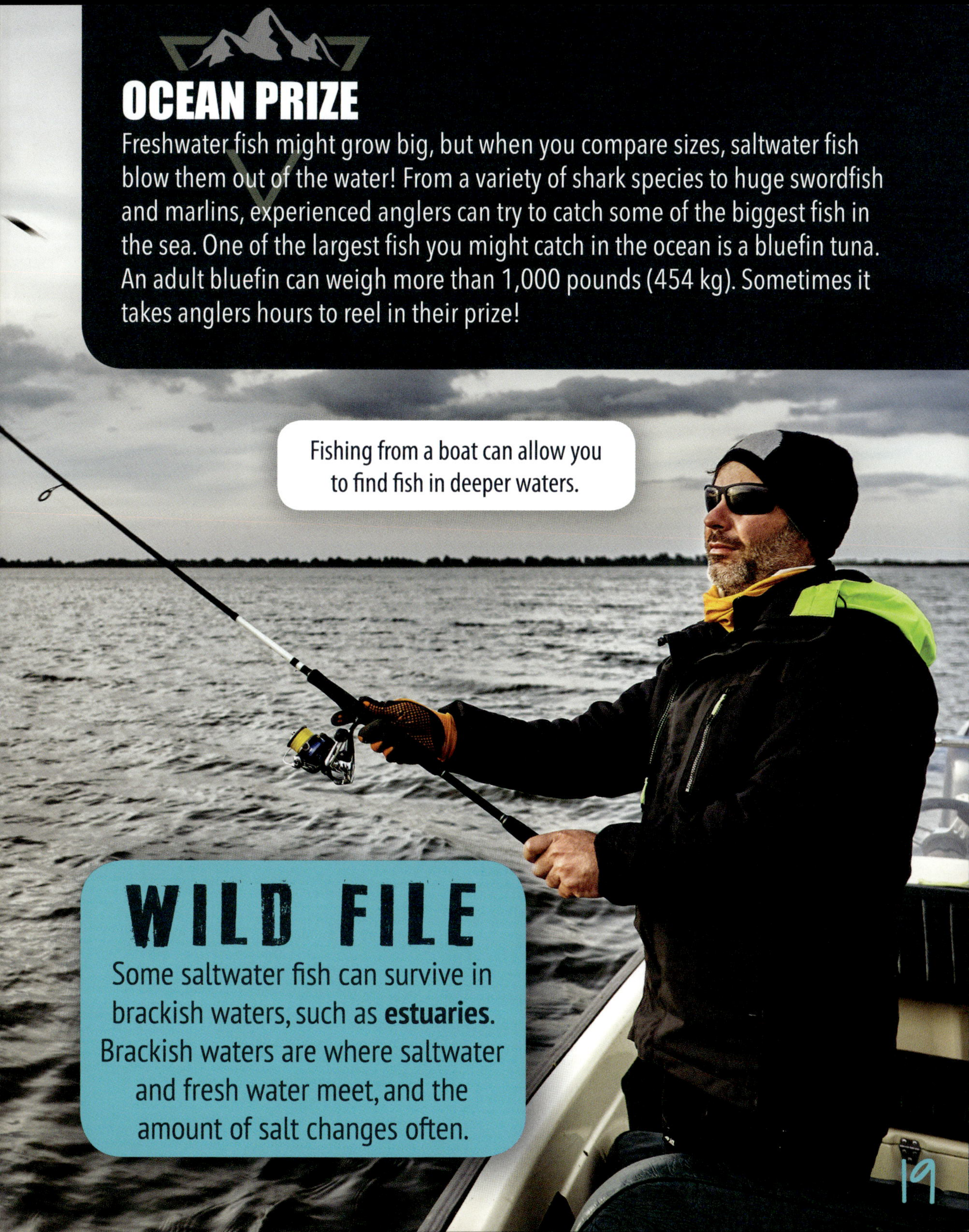

Fishing from a boat can allow you to find fish in deeper waters.

WILD FILE

Some saltwater fish can survive in brackish waters, such as **estuaries**. Brackish waters are where saltwater and fresh water meet, and the amount of salt changes often.

Fly-Fishing Fun

Some anglers use a fly to attract their fish. A fly is bait made of different materials like feathers, hair, or thread. Fish can't see a hook inside human-made flies.

You can fly-fish in rivers, creeks, streams, ponds, and lakes. To begin you need a special fly rod and reel, a fly line, a leader, and a fly. An experienced fly-fisher can cast a line to an exact location. The fly will appear to dance and move on the surface of the water, making it look like a tasty insect. Two types of flies are wet and dry. Wet flies look like young insects living below the water's surface, while dry flies look like adult insects sitting on top of the water's surface.

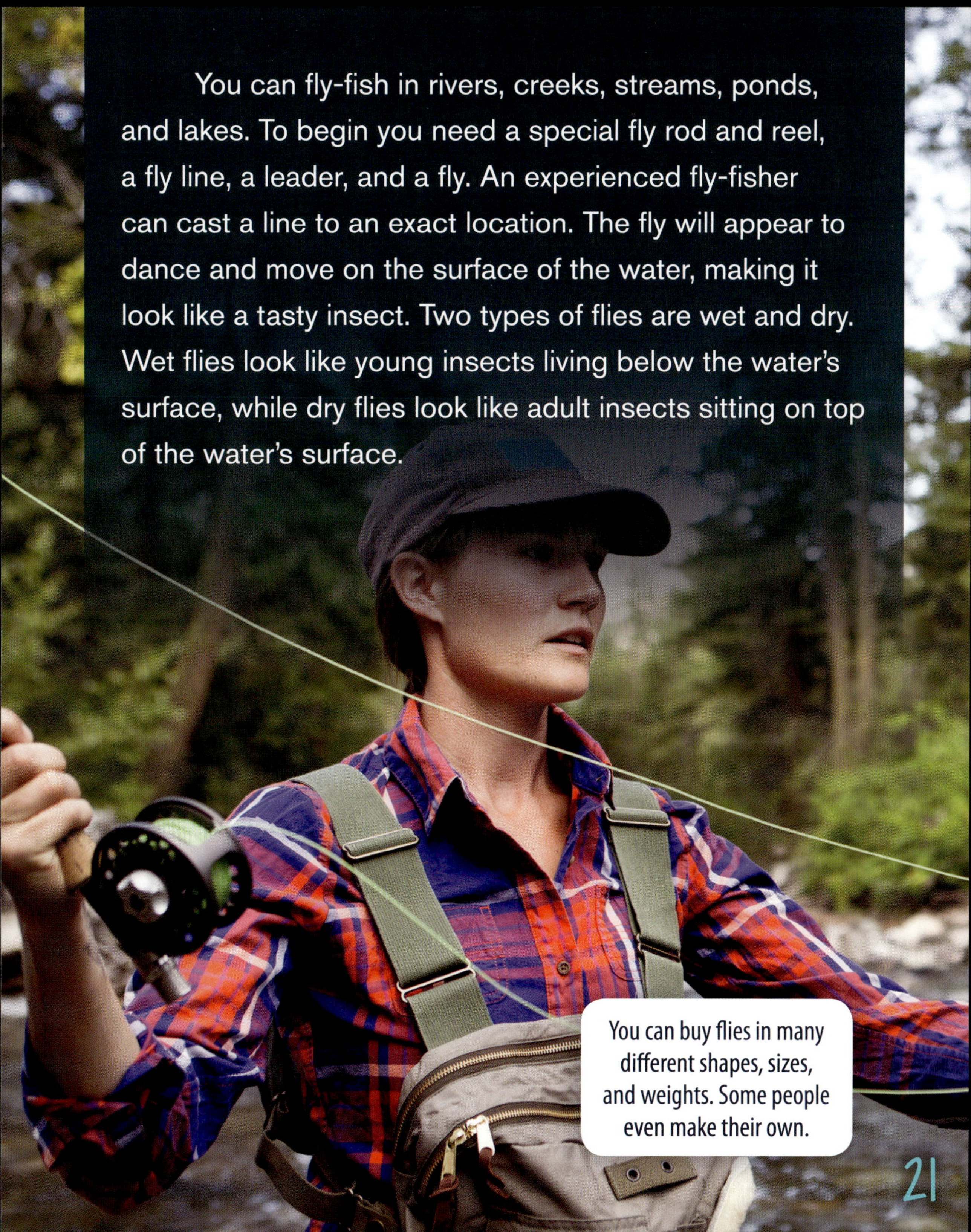

You can buy flies in many different shapes, sizes, and weights. Some people even make their own.

Breaking the Ice

If you live in an area with cold winters, you might want to try ice fishing. This form of fishing comes from native peoples of cold northern regions, like Canada, Greenland, and the northern United States. Ice fishermen break the ice to fish from a hole in a frozen body of water.

You can ice fish when it has been very cold for long periods of time. Ice must be at least 4 inches (10.2 cm) thick before it is safe to walk and fish on. You can use a metal bar with a blade, called a spud, to cut holes in ice. Holes are often more than 8 inches (20.3 cm) across. You should use a shorter rod for ice fishing so you can sit close to your hole.

WILD FILE

For thick ice, anglers may use an **auger**. Some augers are hand tools, while others are powered.

ICE FISHING IN MINNESOTA

Ice fishing is a favorite pastime during a long Minnesota winter. Bodies of water often freeze over by January or February, and that's when the ice fishing can begin! People can drive onto frozen lakes, set up huts and lodges, and camp out while they fish. Some fishing spots turn into winter villages, with people flocking to the frozen water for a chance at a winter catch.

Ice fishing can be one very cold activity! Make sure to dress in warm layers.

what kinds of fish live in a certain body of water. Some fish are caught for sport, while others are caught for food.

WILD FILE

Eels are long fish that have short fins running along their backs. These fish have no scales and can be found in saltwater and fresh water.

People often eat freshwater fish such as perch, bluegill, bass, trout, and salmon. Perch are round, narrow fish that prefer shady areas and often swim in groups. Bluegills are wider, colorful fish that prefer shallow water. Bass are large, brownish fish that prefer deeper water, and they're often attracted by an **artificial** worm or lure. Fishermen find trout and salmon in moving waters, such as rivers. After an angler makes their catch, they can clean and prepare their fish for cooking.

In the U.S. Southeast, grouper is a popular catch. Because of that, it's often on the menu in restaurants in this region.

Pollution and Population

Everything in nature exists in natural **biomes**. Fish depend on other creatures, plants, and conditions to live. That's why it's so important to keep waterways clean and protect fish populations.

In certain areas of the world, pollution causes some fish to be unsafe to eat and others to die. Some places are overfished, which means there aren't enough fish to replace those that are caught. **Conservationists** work to teach people how to keep bodies of water, like lakes and rivers, clean. As overfishing and pollution reduce fish populations, workers may stock some lakes with more fish. Fisheries, or fish farms, have large tanks for raising fish for people to eat. It's important to keep waterways clean and full of healthy fish.

Learn about the issues facing your local bodies of water. How can you help keep them clean?

Hooked on Fishing!

Fishing is a great sport for people who love to be outdoors, and don't mind working on their patience. You might sit for hours waiting for a bite, but that makes your catch even more exciting.

Anyone can fish with a simple fishing pole and hook, and a fishing **license**, of course. To learn more about the sport, you can practice a new skill or try new gear. Angling for larger fish can be quite a challenge requiring strength and patience. Some people test their skills in a fishing tournament, in which a person who catches the largest fish wins a prize. While prizes are nice, many anglers will tell you that there's no better prize than reeling in your fish. Give fishing a try—you might just get hooked!

SAFETY TIPS

- Always have adult supervision when you go fishing.
- Wear a life jacket while fishing—even if you're on a dock. Big fish can pull you in!
- If water looks polluted, choose a new fishing spot.
- Look out for trees, people, and electrical lines before you cast.
- Watch out for teeth or sharp body parts on your fish once you catch it.
- Be careful when using sharp tools to catch or clean your fish.
- Don't fish during lightning storms or other **hazardous** weather.

Fishing can be a fun activity to try when you travel. You can learn about different bodies of water and the fish that call them home.

GLOSSARY

artificial: Made by people.

attract: To cause something to go toward something else.

auger: A tool used for boring holes, as in wood, soil, or ice.

biome: A natural community of plants and animals, such as a forest or desert.

conservationist: A person concerned with conservation, or the care of nature.

designated: Set apart for a specific purpose.

estuary: An area where the ocean's tide meets a river.

hazardous: Involving risk.

license: An official paper giving someone the right to do something.

mimic: To copy the way something looks, acts, or sounds.

prehistoric: Relating to a time before written history.

sonar: Technology that uses sound waves to find things in a body of water.

vibration: A rapid movement back and forth.

FOR MORE INFORMATION

Books

Hogen, Zeb, and Kathleen Weidner Zoehfeld. *Monster Fish! True Stories of Adventures With Animals.* Washington, DC: National Geographic, 2017.

Palmer, Andrea. *We're Going Freshwater Fishing*. New York, NY: PowerKids Press, 2017.

Websites

Ice Fishing Facts for Kids
kids.kiddle.co/Ice_fishing
Explore fun facts about ice fishing!

Freshwater Fish
www.nationalgeographic.com/animals/fish/group/freshwater-fish/
Learn more about freshwater fish with National Geographic.

INDEX